Activities Booklet for Liberal Arts Mathematics and Quantitative Reasoning Courses

Prepared by

Scott Barnett
Henry Ford Community College

 CENGAGE

Australia • Brazil • Canada • Mexico • Singapore • United Kingdom • United States

For product information and technology assistance, contact us at **Cengage Customer & Sales Support, 1-800-354-9706 or support.cengage.com.**

For permission to use material from this text or product, submit all requests online at **www.cengage.com/permissions**.

ISBN: 978-0-357-02562-8

Cengage
200 Pier 4 Boulevard
Boston, MA 02210
USA

Cengage is a leading provider of customized learning solutions with employees residing in nearly 40 different countries and sales in more than125 countries around the world. Find your local representative at: **www.cengage.com**.

To learn more about Cengage platforms and services, register or access your online learning solution, or purchase materials for your course, visit **www.cengage.com.**

Printed in the United States of America
Print Number: 06   Print Year: 2023

TABLE OF CONTENTS

1: Reasoning, Estimation, and Problem Solving

Project 1: Check Digits

Overview

A sequence of numbers or characters used for identification of a person or object often contains a check digit. A check digit helps detect errors made in recording the sequence, and it helps detect fraudulent activity. You will work with check-digit standards in this project.

Materials

1. Calculator (scientific or graphing)
2. Two books published in 2007 or later

Procedure

1. In 2007, the International Standard ISBN Agency adopted the ISBN 13 standard for commercial book identification. ISBN stands for *International Standard Book Number*. Research the ISBN 13 standard. Then answer the following.

    a. How long is an ISBN 13, and where does the check digit appear within it?

    b. What computation is required to determine the check digit?

2. Locate the ISBN 13 for two books published in 2007 or later. Show the computations involved in checking that each book's ISBN follows the ISBN 13 standard.

3. Let's investigate the errors that check digits catch.

    a. Suppose that we are missing one of the digits in each of two different ISBN 13 identifiers. Are we able to determine the missing digit regardless of the location within the identifier of the missing digit? Give two examples, with the missing digit in a different location each time, to support your answer.

    b. Suppose that instead of missing a digit, we have transposed (interchanged) two digits. Does the ISBN 13 standard catch the error every time? Give two examples—one in which the transposed digits are next to each other and one in which they are not—to support your answer.

4. Credit card numbers contain check digits.

    a. Research and describe the Luhn Algorithm.

    b. Write down any 16-digit number in groups of four digits at a time so that your number is formatted as a typical credit-card number might be. Then show the computations involved in using the Luhn Algorithm to determine whether your number could actually be a credit-card number. Be sure to state a clear conclusion.

5. The term 'check *digit*' may be somewhat of a misnomer, and the ISBN-13 and credit-card examples may prompt you to think that the location of the check digit is always the same. Find and describe at least one example of a 'check digit' scheme in which at least one of the following is true:

   a. the 'digit' may be a letter;
   b. the 'digit' is actually a two-digit sequence; or
   c. the digit considered the check digit is located at a position different from the position at which the ISBN 13 or credit-card check digit is located.

Mini-Project 1a: Logic Puzzle

Your textbook may contain logic puzzles for you to solve. If not, you may do an Internet search for logic puzzles to see examples.

Construct a logic puzzle that requires the solver to match four people or items with their corresponding properties (their jobs, for instance, in the case of people, or their retail prices, for instance, in the case of items).

Some tips:

1. Start with the answer. Be sure that each person name or item name corresponds to a separate property.
2. Avoid stating any property directly. ("John is the candlestick maker" makes for a poor clue.)
3. Fill in the blank grid below with your person or item names on the left and their corresponding properties across the top. Use the grid and your clues to solve your puzzle. Eliminate any extraneous clues—clues that are unnecessary due to the presence of other clues.

Mini-Project 1b: Estimation

Find and make a copy of a picture that shows people or items of which it would be time-consuming or impractical to determine an accurate count.

Determine a method of estimating the number of people or items shown in your picture. Then carry out your method.

Determine your estimate from the picture alone, not from background or outside knowledge. For instance, your knowledge of the real-life size of a jar pictured or of the number of packages of jellybeans that may be emptied into it is not to be used in your estimation. However, background or outside knowledge may be used to determine the effectiveness of your estimation method.

2: Sets and Logic

Project 2: Venn Diagrams

<u>Overview</u>

Venn Diagrams may be used to illustrate the possible relationships among sets. You will explore the possible relationships between two sets and among three or four sets in this project.

<u>Materials</u>

- Calculator (scientific or graphing)

<u>Procedure</u>

1. Suppose that every entity in the universal set or universe (U) may be classified according to whether it is *active* (A) or whether it is *big* (B). Then there are four distinct states that a given entity may be in at a given time. One of these states is *active and big*. Using set notation, this may be denoted $A \cap B$.

    a. Construct a Venn Diagram whose regions represent the four states that a given entity may be in. In addition to labels of A and B in your diagram, number each region 1 through 4.

    b. Use set notation to write the remaining three states, and describe those states in words. Each time, indicate which numbered region corresponds to your notation and word description.

2. Suppose that a third classification category is to be added: *catchy* (*C*). Then one of the states that an entity may be in is *active, but neither big nor catchy*. Using set notation and intersections only, this may be denoted $(A \cap B') \cap C'$.

   a. Construct a Venn Diagram whose regions represent all possible states that a given entity may be in according to these three categories. In addition to labels of *A*, *B*, and *C* in your diagram, number the regions of your diagram starting with 1. How many distinct states are there?

   b. Use set notation (intersections only) to write the remaining states, and describe those states in words. Each time, indicate which numbered region corresponds to your notation and word description.

3. Many texts include Venn diagrams that illustrate the possible relationships among two or three sets. Relationships among more sets may be illustrated using Venn diagrams, though. Research Venn diagrams to learn how a four-set Venn diagram may be constructed.

   a. Add a fourth classification category—say, one starting with d and named as set D. Construct a Venn diagram illustrating the relationships among the four categories. In addition to labels of A, B, C, and D in your diagram, number the regions of your diagram starting with 1. How many distinct states are there?

   b. Use set notation (intersections only) to write the states, and describe those states in words. Each time, indicate which numbered region corresponds to your notation and word description.

7

4. The example state in **Question 2** was denoted using intersections as $(A \cap B') \cap C'$. That state may also be denoted $A \cap (B \cup C)'$.

   a. Select two more of the regions in your three-category Venn Diagram, and write alternate set-notation descriptions for them.

   b. Select three regions of your four-category Venn Diagram from **Question 3** and write alternate set-notation descriptions for them.

5. In **Question 3**, your diagram contains shapes other than circles. Research and explain why shapes other than circles must be used in a Venn Diagram that shows the possible relationships among four or more sets.

Mini-Project 2a: Divisibility

For multiple events that each occur with a unique frequency, it may be helpful to determine how often at least one of the events will occur. For example, if a building must be open every seventh day for an event of one type and every eighth day for an event of another, it may be helpful to know on how many times over a certain period the building will need custodial services.

1. Choose a and b, two whole numbers that are relatively prime, meaning that the numbers share no common prime factors. State your numbers. Determine how many numbers between 1 and 1000, inclusive, are divisible by at least one of your numbers.

2. Choose c and d, two composite numbers that share exactly one prime factor and are such that neither divides into the other without a remainder. State your numbers. Again, determine how many numbers between 1 and 1000, inclusive, are divisible by at least one of your numbers.

3. Choose r and s, two unequal whole numbers that are such that one number is divisible by the other. State your numbers. Yet again, determine how many numbers between 1 and 1000, inclusive, are divisible by at least one of your numbers.

4. Research the Inclusion-Exclusion Principle. How did that principle enter into your computations in **Questions 1** through **3**? If your method did not involve this principle, then how might it have?

9

Mini-Project 2b: Formal Logic

Given here are two valid arguments stated symbolically.

In each case, write mathematical or non-mathematical statements to illustrate these arguments such that all statements involved in the arguments are true.

As an example:

$r \rightarrow s$

$\sim s$

$\therefore \sim r$

All whole numbers divisible by 10 end in 0. 75 does not end in 0. Therefore, 75 is not divisible by 10. **Or:**

Every fruit contains seeds. A pencil does not contain seeds. Therefore, a pencil is not a fruit.

1.
$$r \vee s \rightarrow t$$
$$t \rightarrow \sim v$$
$$\underline{v \qquad\qquad}$$
$$\therefore \quad \sim r$$

2.
$$\sim m \rightarrow n$$
$$n \rightarrow p \wedge q$$
$$\underline{\sim q \qquad\quad}$$
$$\therefore \quad m$$

3: Proportions and Variation

Project 3: Scale Drawing

<u>Overview</u>

A scale drawing is a diagram that represents an entity such as a building or a geographic region such that corresponding lengths are in proportion. You will construct one type of scale drawing, a floor plan, in this project.

<u>Materials</u>

- Calculator (scientific or graphing)
- Measuring tape and ruler, each with both U.S. Customary and metric units
- Graph paper

<u>Procedure</u>

1. Choose a structure with which you are familiar and to which you have ready access. Decide whether you will draw a floor plan of the entire structure or just a portion. A house of 1000 to 2000 square feet on one or two levels may work well.
   a. Outline the walls (exterior and interior) of the structure you've chosen. Do not be concerned with correct proportions at this point; you are simply using this initial diagram as a place to record measurements.

b. Measure the true lengths of the walls. Record the true lengths in both U.S. Customary and metric units along the corresponding lengths in **Part a**.

2. Compute the proper lengths of the floor-plan-diagram line segments (or perhaps curves) that will represent your structure's walls.

    a. One diagram will use U.S. Customary units. Choose a scale that allows your diagram to fit in the **Part b** space below. An example of a scale might be "1/8 inch to 1 foot." This means that for every length of 1 foot in the structure there will be a corresponding line segment or curve drawn on graph paper that has length 1/8 inch.

        Record your chosen scale here: _____

        Show the calculations involved in determining floor-plan lengths for the diagram that will use U.S. Customary units.

    b. Draw your diagram here:

    c. One diagram will use metric units. Choose a scale that allows your diagram to fit in the **Part d** space below.

    Record your chosen scale here: _____

    Show the calculations involved in determining floor-plan lengths for this metric-unit diagram.

    d. Draw your diagram here:

3. Diagramming a floor plan comes with challenges, as you've likely found in this exercise.

    a. Examine the lengths of the walls of your chosen structure. How does the sum of the interior lengths along one side compare with the exterior length along that side? If the interior-length sum you've determined does not equal the exterior length, what might account for the difference?

    b. Aside from wall lengths, what measurements might it be desirable to document in your floor plan? (Give at least two.) What challenges are involved in adding those measurements to a diagram?

4. Professional blueprints are created using computer-aided design (CAD) software. Research CAD software. What mathematical concepts must be understood in order to effectively use CAD software? (Give at least three.)

Mini-Project 3a: Scaling a Recipe

Find a recipe that requires at least four ingredients and makes four servings.

1. Scale the ingredient amounts so that there will be ten servings instead of four. Show the computations required.

2. Scale the ingredient amounts so that there will be fifteen servings instead of four. Do this by
   a. starting with the four-serving recipe and

   b. starting with the ten-serving recipe

3. Are your answers in **Parts a** and **b** of **Question 2** equal? Why or why not?

4. Does everything (not just ingredients) involved with the recipe scale proportionately when the number of servings increases?

Mini-Project 3b: Inverse Variation

For a given distance traveled, the rate and time involved vary *inversely*: as distance is the product of rate and time, an increase in the rate, for instance, results in a decrease in the time.

1. Find at least two other examples of quantities that vary inversely. Write the mathematical formulas giving these inverse-variation relationships and state the meanings of the variables involved.

2. For each example, what happens to one variable when the other is doubled? Halved? Multiplied by an arbitrary positive value k?

3. Occasionally the product of two quantities will be a *negative* constant. For example, in algebra, the product of the slopes of two nonvertical perpendicular lines is always −1. When the product of two quantities is negative, is it correct to say that an increase in one always results in a decrease in the other? Discuss.

4: Finances

Project 4: Credit Cards

<u>Overview</u>

Credit cards allow the holder to make purchases without cash or when funds are not immediately available. The holder promises to pay at a future time. In exchange for this convenience and this trust, interest—an amount paid at a particular rate for borrowed money—is generally charged. You will explore the lengths of time needed to pay off a credit card in this project.

<u>Materials</u>

- Calculator (scientific or graphing)

<u>Procedure</u>

1. Suppose that you have a balance of $3000 on a credit card. (You may wish to instead use an actual balance from a credit card you hold.) Assume that you will make no new charges to the card.

   a. With an Annual Percentage Rate (APR) of 22.9% and a minimum monthly payment of 5% of the balance, how long will it take to bring the balance down to $1?

      Write any relevant formulas here:

      Show your computations here:

   b. Research current typical APRs and minimum-monthly-payment requirements, or use those values from a credit card you hold. Again, assume that you will make no new charges to the card. How long will it take to bring the balance down to $1? (Show your computations.)

18

2. Start with the same balance as you began with in **Question 1**. Suppose now that each month you will pay a fixed percentage *of the initial balance*.

   a. Suppose that there is a 22.9% APR, and suppose that you will pay 5% *of the initial balance* each month. How long will it take for the card to be paid off?

   Compute 5% of the initial balance here:

   Show the computations involved in determining the number of months until payoff here: (You may need to use a spreadsheet or attach separate paper. Show at least the first two months' computations here.)

   b. Suppose that the APR used in **Question 1**, **Part b** is in effect, and suppose that you will pay 5% *of the initial balance* each month. How long will it take for the card to be paid off? (You may need to use a spreadsheet or attach separate paper. Show at least the first two month's computations here.)

3. Suppose that there is a rule for the credit cards here that requires a monthly payment of 5% of the current balance or $20—whichever is greater. Determine the specific effect, if any, that this rule has on your answers to **Questions 1** and **2**.

4. Credit-card companies offer various promotional offers to encourage credit-card use. Find an example of at least one such promotional offer, and determine the effect that it would have on the card-payoff duration you determined for **Part b** of **Question 1**.

Mini-Project 4a: Car Loan

Find an advertisement for a car that you would like to purchase. Also, research current interest rates for a 48-month loan for the car. (Note that your interest rate will typically depend on whether the car of interest is new or used.)

An *amortization table* outlines the schedule of payments resulting in loan payoff. A typical amortization table shows, for each payment, the date, the total payment amount, the amounts applied to principal and interest, and the amount still due. A typical header row may be as follows:

Date	Amt. of Payment	Amt. to Interest	Amt. to Principal	Loan Balance

Construct an amortization table for a loan for the advertised car at the interest rate you've researched. Include outside of the table a statement of the interest rate applied.

Mini-Project 4b: Retirement

Many people expect to leave the workforce permanently at some point in their lives. This period, known as retirement, still has life's expenses associated with it!

Do the following:

1. Research the amount of money it typically takes to live one year in retirement. Actual amounts vary incredibly widely based on a multitude of factors, but still: 'average' amounts may be calculated.
2. Research current long-term savings rates. A typical set of retirement accounts, known as a *portfolio*, will generally contain more than just a savings account.
3. Determine the number of years you expect to be in the workforce prior to retirement, and estimate the number of years you expect to live in retirement. (You may wish to research life-expectancy in order to help you.)

Use these values to calculate the amount of money you'll need to save each month in order to finance your retirement assuming that no other income sources (such as Social Security in the United States or an employer pension) are available to you. Show the calculations involved.

5: Linear Functions

Project 5: Scaling Linear Models

Linear models fit many life situations. Shipping items to a customer may cost one amount for a box and the shipment itself, plus another amount per ounce or per pound. Some professionals may charge an acceptance or intake fee for their services, plus a certain amount per hour or session.

In this project you'll set up a linear model and determine the effects various adjustments have on the model.

Materials

- Calculator (scientific or graphing)

Procedure

1. Research a price or cost that may be modeled linearly. What is the base rate? What is the per-unit rate?

2. Write the relevant linear function. Be sure to define your variables.

3. Use the linear function written in **Question 2** to calculate the cost of two units of the independent-variable quantity. State that cost.

4. How does the linear function rule change if the base rate remains as stated in **Question 1** but the cost of two units will have each relationship given here to the cost calculated in **Question 3**?
    a. The cost of two units is doubled.

    b. The cost of two units is halved.

    c. The cost of two units is multiplied by an arbitrary positive constant k.

5. How does the linear function rule change if the base rate remains as stated in **Question 1** but the number of units obtained for the price determined in **Question 3** is now
    a. halved (so that just one unit has the price determined in **Question 3**);

    b. doubled (so that for units are obtained for the price determined in **Question 3**);

    c. multiplied by an arbitrary positive constant k (so that $2k$ units are obtained for the price determined in **Question 3**)?

6. How does the linear function rule change if both the base rate and the cost of two units will have each relationship given here to the original base rate and to the cost of two units calculated in **Question 3**, respectively?

    a. Both costs are doubled.

    b. Both costs are halved.

    c. Both costs are multiplied by an arbitrary positive constant k.

7. Did you find one of **Questions 4**, **5**, or **6** easier to answer than the other two? If so, which one—and why?

Mini-Project 5a: Service Rates

Suppose that you've set up a business selling non-specialized home services (light carpentry, cleaning, and the like) for two hours to those who live in your region. You've determined that the market bears a price of $150 for this service.

Suppose you've now determined that (1) to discourage cancelled appointments, some amount of the $150 needs to be considered a non-refundable charge and (2) you wish to offer a variety of appointment durations (total time lengths) and increments—specifically: quarter-hour increments will be acceptable.

Decide a non-refundable amount you will charge per appointment. Then construct a function that takes as its input the number of quarter-hours an appointment takes and gives as its output the price. The function you construct must be consistent with a two-hour appointment having price $150.

Vary the non-refundable amount. What effect does the size of the non-refundable amount have on total price as job-duration varies?

Mini-Project 5b: Linear Regression and Correlation

Find at least five data points for two measurement variables you believe may be linearly correlated. You may use data you already have—your chapter homework average vs. corresponding chapter-test scores in a particular class, for example—or data available from external sources.

1. Choose one of your variables to be considered the independent variable. With this variable as the independent variable, use a calculator or a spreadsheet to compute the linear-regression equation and correlation coefficient for your variables.
2. Does the correlation coefficient support your initial belief that the measurement variables are linearly correlated? Discuss.
3. Redo the computations from **Question 1** with the other variable as the independent variable. Does your conclusion from **Question 2** change? Why or why not?

6: Nonlinear Functions

Project 6: Population Growth

Over limited time periods the growth of living populations—whether the populations under consideration are the people in a region or the bacteria in a petri dish—may be modeled with exponential functions. Over longer time periods, though, exponential functions may be inappropriate.

In this project, you'll construct and examine the limitations of an exponential function. You'll then examine another type of function that may be more suitable when an exponential function is not.

<u>Materials</u>

- Calculator (scientific or graphing)

<u>Procedure</u>

1. Find a population value at three different times—with at least the first two values being the populations at consecutive time units. Record your three data points here.

2. Determine the growth rate of the population between the first two time units using your data. Use this growth rate and the first data point to construct an exponential model for this population. (Consider the time of the first population value as time 0.)

3. Use the exponential model constructed in **Question 2** to estimate the population after one time unit and after the number of time units from your third data point in **Question 1**. How do the populations computed using your model compare to the respective known populations (from your data) in each case?

4. Use the exponential model constructed in **Question 2** to estimate the population after 10, 100, and 1000 time units. How reasonable are your results in each case?

5. Research *logistic* functions. Use a calculator or spreadsheet and the three data points from **Question 1** to determine a logistic model for this population. (Again, consider the time of the first population value as time 0.)

6. Use the logistic model from **Question 5** to estimate the population after 10, 100, and 1000 time units. How reasonable are your results in each case?

7. Use technology to graph the exponential model from **Question 2** and the logistic model from **Question 5** on the same axes. Compare the graphs of the two models, describing any similarities or differences you notice.

8. The graph of your logistic model should illustrate a *carrying capacity*.
   a. What is a carrying capacity?

   b. What is the carrying capacity for your model?

   c. How does the graph illustrate this carrying capacity?

Mini-Project 6a: Logarithmic Functions

Logarithmic models describe many physical quantities. Acidity, sound intensity, and earthquake intensity are among the quantities typically described using logarithmic models. Let's consider sound intensity.

1. Research the sound intensity of a typical vacuum cleaner. What are the measurement units for sound intensity? To what decibel level does this intensity correspond?

2. Suppose that two of these typical vacuum cleaners are running side by side. Using your answer to **Question 1**, what would the sound intensity be? To what decibel level does this intensity correspond?

3. How many vacuum cleaners must be in operation to double the decibel level? Is this desirable? Practical?

4. Research the human pain threshold for sound. How many vacuum cleaners would it take to reach this threshold?

5. Determine each:
   a. how sound intensity is affected when the decibel level is multiplied by an arbitrary positive constant k and

   b. how decibel level is affected when sound intensity is multiplied by an arbitrary positive constant k.

Mini-Project 6b: Maximizing Revenue

Suppose that you have acquired a comfortable bus and a chauffer's license. You will operate trips to a city some distance away. You are interested in maximizing the revenue—the amount of money you take in—from these trips.

Suppose that you've determined that you can sell 30 seats at a cost of $200 each. This gives you a revenue of $30 \times \$200,$ or $6000.

1. Demand for your trips may depend in part on the price you charge for your trips. Estimate the amount of a price decrease that would be necessary to prompt one more person to take your trip. Then calculate the revenue generated. Does your revenue increase?

2. Assume that every time the amount decreases by the amount you gave in **Part a**, one more person will take your trip. Write a function that models the revenue generated when x price decreases are implemented. What would a negative value of x mean in this case?

3. The function in **Part c** is quadratic (or at least should be!). Determine the value of x that maximizes revenue. What is the maximum revenue?

4. Why might maximizing profit be more of interest than maximizing revenue? What other information would be needed, and how might your work be different, if you were maximizing profit?

7: Introduction to Probability

Project 7: Games of Chance – Casinos

<u>Overview</u>

While you may hope for a big win at a casino, a casino on average will make money on its games. In this project you will design a casino card game using the concept of mathematical expectation to yield a game that you believe gamblers will want to play—but that will allow the 'house' (the casino itself) to earn its take.

<u>Materials</u>

- Calculator (scientific or graphing)

<u>Procedure</u>

1. Research the playing cards in a standard 52-card deck. Categorize the cards by suit, color, and value, stating the cards in each category and counting their number.

2. Calculate the probabilities of each of the following:
   a. selecting a face card when a card is drawn at random from a standard deck

   b. selecting a face card followed by a non-face card when two cards are drawn from a standard deck sequentially, without replacing the first card before drawing the second

   c. selecting a face card and a non-face card (without regard to order of selection) when two cards are drawn from a standard deck without replacement

   d. selecting a face card followed by a non-face card when two cards are drawn from a standard deck sequentially, replacing the first card before drawing the second

31

3. Calculate the mathematical expectations for a player for each event in **Question 2** if in each case a player pays \$2 in order to have the card(s) drawn and wins \$20 if the indicated event occurs.

4. Questions 2 and 3 are designed to give you some familiarity with expectations resulting from betting on card events. Now, name and design a game that meets the following conditions:

    a. the game is based on use of a shuffled 52-card deck

    b. the game specifies the number of cards (two or more) to be selected for each hand, and it specifies whether these cards are selected with or without replacement

    c. the game specifies at least six different hands that are considered winning hands

    d. the game specifies at least two different amounts that may be bet per hand, and it designates corresponding prize amounts for winning hands

    e. the casino has a mathematical expectation of between 30% and 70%, inclusive, of the amount bet.

Construct a table indicating the per-hand payoff for each bet and winning hand here:

5. Decks of playing cards often come with two *joker* cards. A joker card is not part of any suit and may be considered good or bad to have in a hand, depending on the game, if the joker cards are played at all. Determine new expectations for the game you designed in **Question 4** if:

    a. joker cards are played and are considered 'wild' (may be counted as any of the desired cards in the hands designated as winning hands)

    b. joker cards are played but never count as any of the desired cards.

Mini-Project 7a: Games of Chance – Lotteries

Most states in the Unites States and provinces in Canada have lotteries. Lottery players bet that a certain number or combination of numbers will be drawn on a given day and win amounts that may depend on a combination of the bet, the number of drawn numbers matched, and the number of other players.

Research a state or provincial lottery in which between five and seven numbers (inclusive) are drawn from 45 to 55 numbers (inclusive) and answer the following:

1. What are ticket prices (there may be more than one for a single state), and what is the minimum number of numbers that must be matched in order to win for each ticket price?

2. For each ticket price and winning number of numbers, state the amount (or typical amount) won, and the odds or probability displayed on the official state lottery website.

3. Use concepts from your study of odds and probability to show the computations behind the odds or probability displayed in each case on the official state lottery website.

4. Calculate a lottery player's mathematical expectation from the purchase of a single maximum-priced lottery ticket in your chosen state.

Mini-Project 7b: License Plates

Choose a governmental unit, such as a state in the United States or province in Canada, that issues license plates. Research the rules for license-plate formats for standard passenger vehicles in your chosen governmental unit.

1. How many non-space characters are used? What are the letter and digit restrictions? (For purposes of this mini-project, ignore any 'bad word' restrictions.)

2. Calculate the number of different plates that are possible given your responses in **Question 1**.

3. Suppose that any letters in your first name that are not already eliminated from license plates in your chosen state are now eliminated. How many license plates are possible now? Show all calculations.

4. Suppose that the location of a space matters such that, for example, ABC 123 is a different plate than AB C123. How many additional plates beyond the number calculated in **Question 2** are possible if the restrictions from **Question 1** (only) apply and the license plate is to contain a total of one space—but between any two characters?

35

8: Statistics

Project 8: Snack-Food Consistency

<u>Overview</u>

What is typically in a snack bag of your favorite candy, trail mix, or other go-to food? In this project, you'll investigate.

<u>Materials</u>

1. Calculator (scientific or graphing)
2. At least 10 same-size snack bags of a candy, mix, or such that may be separated into at least three categories (e.g. at least three colors for candy; into peanuts, raisins, and crackers for trail mix)

   Be sure to record all values needed before consuming your data sources.

<u>Procedure</u>

1. Decide the three (or more) categories into which you'll separate your snack bags.

2. Open each snack bag and separate its contents according to your categories. Record the number of individual snack-bag items that fall into each category.

3. Determine the minimum and maximum number of items in each category across your bags, and determine the quartiles.

4. Determine the mean and the standard deviation of the number of items in each category across your bags. What percent of the data in each category fall within one standard deviation of the mean? Two standard deviations?

5. Construct a summary in which your results from **Question 3** are presented in a box-and-whisker plot and your results from **Question 4** are presented in tabular format.

Mini-Project 8a: Approximately Normal Distributions

Many real-life data distributions are approximately normal.

Find a set of data containing 100 or more data points and related to a topic you're interested in. You might examine the finish times in a particular marathon, the salaries of the teachers in a local public-school district, or other data generally readily available online.

1. Use spreadsheet software to display a histogram of your data. Does the histogram appear to approximate a normal distribution in shape?

2. Calculate the mean and standard deviation for your data set and display the results with your histogram. What percentage of your data set falls within one standard deviation of the mean?

3. How does that compare with the usual percentage of a normally distributed data set falling within one standard deviation of the mean?

Mini-Project 8b: Grade-Point Average

Research your institution's grading scale. Throughout this project, references to grade points may be replaced with references to quality points or similar points depending on your institution's nomenclature.

1. What is the highest grade worth numerically? The lowest? Construct a table that shows for each grade the number of grade points that the grade is worth.

2. Most institutions use a weighted average to compute a grade-point average. Assuming that yours does, show the computations that lead to your current GPA. If you are a first-semester student taking more than one class, estimate your expected GPA by computing an appropriate weighted average. If you are a first-semester student taking only one class, or if you have earned the same grade in every course to date, calculate the cumulative GPA that would result from variation in grades over four or five upcoming courses.

3. What effect would earning the highest grade your institution offers in a three-credit or three-unit course have on a student who after completion of 20 credits or units has the numerical equivalent of a 'C' (or other middle-grade) average. After completion of 50 credits or units? After which number of credits/units does the greater effect occur?

9: Voting and Apportionment

Project 9: Arrow's Impossibility Theorem: We Can't Always be Fair

Overview

It has been proved that when three or more choices are involved, it is not possible to design a voting system that meets these criteria simultaneously:

1. If any choice receives more than 50% of first-place votes, then that choice wins.
2. If a choice wins, then that choice still wins if other voters change their preferences to support it.
3. If a choice wins each time it is positioned one-on-one against every other option, then the choice wins overall.
4. If any choice other than the winner is withdrawn from the race, then the winner does not change.

This impossibility is known as *Arrow's Impossibility Theorem*. Note that the theorem does *not* mean that every election among three or more candidates in every system violates at least one of these criteria. To take an extreme example: with plurality voting (in which the candidate chosen first most often wins), if Candidates A, B, and C each run, and if the candidates are ranked from best to worst as A, B, and C by 1000 of 1000 eligible voters, then Candidate A is the winner, and none of the four criteria are violated.

Materials

*   Calculator (scientific or graphing)

Procedure

1. Suppose that 100 votes in total are cast for options A, B, and C. Votes fall as follows:
    45 voters rank the candidates C (first), then B, then A
    40 voters rank the candidates B (first), then A, then C
    15 voters rank the candidates A (first), then B, then C

    a. Research the plurality method if your course has not covered this topic. Describe the method.

    b. Which candidate wins using the plurality method?

    c. Are any of the four criteria above violated by this method in this election? If so, which one(s), and why?

40

2. Choose a city with a population of at least 1000.
Record the population of the city here: _____

Suppose that 20% of the population will vote in an upcoming mayoral election. To the nearest whole number, how many people will vote?

For Questions 3 through 5, you will assume that these voters choose among candidates X, Y, Z in the mayoral election.

3. Research the Borda Count method. Give a possible distribution of candidate rankings (such as that given in **Question 1**) for the voters in your chosen city's mayoral election that illustrates the Borda Count's violation of Criterion 1.

4. Give a possible distribution of candidate rankings for the voters in your chosen city's mayoral election that illustrates the Borda Count's violation of Criterion 4.

5. Research voting systems to learn of an existing system not yet discussed. Then give a possible distribution of candidate rankings for the voters in your chosen city's mayoral election that illustrates the violation of at least one of the four criteria by your chosen system. Which criterion or criteria are violated, and why?

Mini-Project 9a: Fair Division

Suppose that you and two friends will rent a resort villa for a week. The villa will have three bedrooms of varying levels of desirability.

1. Research the typical cost of such a rental. Record the typical cost here: _____

2. Research the *method of sealed bids*. Describe that method here.

3. Create descriptions for each of the bedrooms such that the rooms are likely to vary in desirability among you and your friends.

4. Assign each bedroom its value to you such that the sum of the values assigned equals the total rental cost. Ask two friends to do the same, or estimate the values two friends might assign.

5. Illustrate the method of sealed bids to determine the amounts you and each friend will contribute to the rental rate.

Mini-Project 9b: Apportionment Paradoxes

Suppose that a population initially of 15,000 is split among three regions of unequal populations and has 12 representatives in total. Give names to the three regions, and then do the following:

1. Research the Hamilton Apportionment method and the population paradox. Then give an example of two population distributions, one for each of two consecutive election cycles, that illustrate the population paradox.

2. Research other apportionment paradoxes. Select one, and give an example of it based on the regions you've named and the populations you've assigned.

10: Circuits and Networks

Project 10: Tree Traversal

Overview

A spanning tree is a minimal set of edges of a graph that covers all vertices of the graph. A minimum spanning tree in a weighted graph is a spanning tree with the smallest possible sum of edge weights. In this project, you'll implement two algorithms for finding minimum spanning trees.

Materials

- Calculator (scientific or graphing)

Procedure

1. Select six cities that are far enough apart from each other and from your current location such that air travel between any pair makes sense. Research typical one-way airfare between each pair of cities. (Assume that a ticket between two cities costs the same regardless of the city in which the trip begins.)

2. Construct a graph in which the cities from **Question 1** are the vertices and the edges are weighted by the airfares.

Suppose that six illustrators—one in each city—are hired to illustrate a book you've written. A draft of the book will be printed in the city in which the first illustrator to work on it dwells and will be shipped to the publisher from the city in which the sixth illustrator dwells. The publisher finds your need to be with the book as it is illustrated in each city excessive. It has agreed to fund your travel to the first illustrator and home from the sixth, but it insists that you pay your own airfare between the individual illustrators.

You are interested in visiting each city exactly once to meet with the illustrator in that city. The order in which you work with the illustrators does not matter to you. Since you are funding the travel from one illustrator to the next yourself, you are interested in minimizing its cost. You are, therefore, interested in a minimum spanning tree of the graph you constructed in **Question 2**.

3. Research *Prim's Algorithm* for finding a minimum spanning tree.
   a. Outline the steps of this algorithm.

   b. Apply this algorithm to determine a minimum spanning tree for the graph constructed in **Question 2.**

4. Research *Kruskal's Algorithm* for finding a minimum spanning tree.

    a. Outline the steps of this algorithm.

    b. Apply this algorithm to determine a minimum spanning tree for the graph constructed in **Question 2.**

5. Do your answers in **Questions 5** and **6** agree with each other? In what order(s) should you travel from one illustrator to the next in order to minimize your self-funded-airfare costs?

Mini-Project 10a: Scheduling

Suppose that the seven student clubs at a school are to hold their organizational meetings on the same day. Some students are members of multiple clubs, and meetings are to be scheduled so that two clubs that share at least one member do not meet simultaneously.

Suppose that there are nine pairs of clubs that share members.

1. Give names to the seven clubs, and select nine pairs of clubs that will have shared members.

2. In the grid below, label the names of the clubs on the top and on the left, and mark an X in the eighteen boxes corresponding to club pairs with shared members.

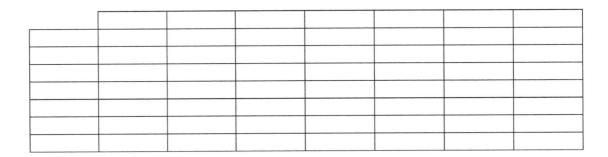

3. Use graph coloring to determine the minimum number of time slots necessary to ensure that no member in two or more clubs has to miss the meeting of one club due to a conflict with the meeting of another.

Mini-Project 10b: Is Greed Always Best?

Select four cities that are far enough apart from each other such that air travel between any pair makes sense. Research typical one-way airfare between each pair of cities. (Assume that a ticket between two cities costs the same regardless of the city in which the trip begins.)

Construct a graph in which these cities are the vertices and the edges are weighted by the airfares. Then complete the following:

1. Select a city from which to start. Research and apply the greedy algorithm to determine a reasonable solution to the problem of flying to each city and returning to the city of origin without overspending.

2. Research and apply the edge-picking algorithm (a type of greedy algorithm, except that we begin with a specific edge rather than an arbitrary vertex) to determine a reasonable solution to the problem of flying to each city and returning to the city of origin without overspending.

3. List *all* possible circuits among the four cities, and compute their costs. Did either or both of your answers in **Questions 1** and **2** find the optimal circuit(s)? Explain.

11: Geometry

Project 11: Tessellations (Tilings)

<u>Overview</u>

A regular tessellation (tiling) of a geometric plane is a covering of the plane with copies of a single regular polygon such that there are no gaps and no overlaps. A tessellation is semi-regular if it consists of two or more regular polygons.

In this project you'll examine regular and semi-regular tessellations.

<u>Materials</u>

- Calculator (scientific or graphing)
- Drawing paper
- Ruler or Straightedge and Compass

<u>Procedure</u>

1. Starting with an equilateral triangle, sketch regular polygons with successively increasing numbers of sides. Separate each regular polygon into as many triangles as possible by drawing all diagonals from one vertex. Multiply the number of triangles formed within each polygon by 180, and divide each result by the number of sides. You now know the measure of each angle of the polygon. We are interested in results which are factors of 360.a t After what number of sides can you conclude that there will be no more factors of 360 resulting as angle measures?

2. For each regular polygon that has an angle measure that is a factor of 360, draw a tessellation (tiling). Also, attempt to draw a tessellation with a regular polygon determined not to work in **Question 1**. What happens?

3. Find two semi-regular tessellations each of which uses copies of the same two types of regular polygons (each tessellation a different number of each, though) to cover the plane. Use angle measures to explain your work.

4. Find a semi-regular tessellation that uses copies of the same *three* regular polygons to cover the plane. Use angle measures to explain your work.

5. Research semi-regular tessellations. How many are there in addition to the three that you have drawn in **Questions 3 and 4**? Use angle measures to explain the existence of each additional semi-regular tessellation.

6. For each tessellation in **Question 1**, draw its *dual*—the diagram resulting from drawing a line segment to connect the center of each polygon. What do you notice?

Mini-Project 11a: Areas of Regular Polygons

A regular polygon is one in which all sides have the same length and all angles have equal measure.

An *apothem* of a regular polygon is a segment connecting the center perpendicularly to a side.

1. Draw a regular hexagon. Separate the hexagon into six triangles by drawing line segments from the center to each vertex. How might separating the hexagon in this manner, together with knowledge of the formula for the area of a triangle, allow you to find the area of the hexagon? Use the word *apothem* appropriately in your answer.

2. Repeat the work in **Question 1** for a regular octagon.

3. Generalize your work in **Questions 1 and 2** to determine a formula for the area of any regular polygon of *n* sides.

4. The number *pi* is defined as the ratio of the circumference of any circle to its diameter. As the number of sides of a regular polygon increases, the polygon begins to look more like a circle, and length of the apothem tends toward the length of the radius of that circle. Use this tendency, the formula you developed in **Question 3**, and the definition of *pi* to informally justify the formula $A = \pi r^2$ for the area of a circle.

Mini-Project 11b: Dimensional Factors

What effect does multiplying dimensional measurements of a figure or object by two, three, or an arbitrary constant k have on the area or volume of the resulting figure or object?

1. Select a two-dimensional figure that you're able to calculate the area of, and calculate the area. Then multiply any lengths involved in the area calculation by two. What would the area of the resulting figure be? How does this area compare to the area of the original figure?

2. Repeat **Question 1** by multiplying the original lengths by 3 and then by k, where k is an arbitrary positive constant. Compare the resulting areas to the original area. What rule are you able to write for determining the area of the figure that results from multiplying an original figure's lengths by k?

3. Select a three-dimensional figure that you're able to calculate the volume of, and calculate the volume. Then multiply any lengths involved in the volume calculation by two. What would the volume of the resulting figure be? How does this volume compare to the volume of the original figure?

4. Repeat **Question 3** by multiplying the original lengths by 3 and then by k, where k is an arbitrary positive constant. Compare the resulting volumes to the original volume. What rule are you able to write for determining the area of the figure that results from multiplying an original figure's lengths by k?